S0-CUQ-317

BEASTLY BOOKS FOR THE BRAVE

THIS BOOK CAN POISON!

THERESE M. SHEA

Please visit our website, www.garethstevens.com. For a free color catalog of all our high-quality books, call toll free 1-800-542-2595 or fax 1-877-542-2596.

Cataloging-in-Publishing Data

Names: Shea, Therese M.
Title: This book can poison! / Therese M. Shea.
Description: New York : Gareth Stevens Publishing, 2020. | Series: Beastly books for the brave | Includes glossary and index.
Identifiers: ISBN 9781538233511 (pbk.) | ISBN 9781538233528 (library bound) | ISBN 9781538233542 (6pack)
Subjects: LCSH: Poisonous animals--Juvenile literature. | Venom--Juvenile literature. | Animal defenses--Juvenile literature.
Classification: LCC QL100.S5328 2019 | DDC 591.6'5--dc23

Published in 2020 by
Gareth Stevens Publishing
111 East 14th Street, Suite 349
New York, NY 10003

Designer: Katelyn E. Reynolds
Editor: Kate Light

Photo credits: Cover, p. 1 (frog) Sergio Foto/Shutterstock.com; cover, pp. 1–24 (book cover) Ensuper/Shutterstock.com; cover, pp. 1–24 (tape) Picsfive/Shutterstock.com; cover, pp. 1–24 (decorative elements) cute vector art/Shutterstock.com; cover, pp. 1–24 (book interior and wood background) robert_s/Shutterstock.com; pp. 4–21 (fun fact background) Miloje/Shutterstock.com; p. 5 Mirko Graul/Shutterstock.com; p. 7 (main) Jason Edwards/National Geographic/Getty Images; p. 7 (map) Ridvan EFE/Shutterstock.com; pp. 9, 15, 17 (inset photo frame) Krasovski Dmitri/Shutterstock.com; p. 9 (both) Auscape/UIG via Getty Images; p. 11 Timothy Fadek/Corbis via Getty Images; p. 13 Joao Paulo Burini/Moment Open/Getty Images; p. 15 (main) IrinaK/Shutterstock.com; p. 15 (inset) Brian Basgen/Musides/FSV/Wikipedia.org; p. 17 (main) Vaclav Sebek/Shutterstock.com; p. 17 (inset) Vladislav T. Jirousek/Shutterstock.com; p. 19 Photography by Shin.T/Moment/Getty Images; p. 21 (platypus) worldswildlifewonders/Shutterstock.com; p. 21 (Hawksbill turtle) Rich Carey/Shutterstock.com; p. 21 (hooded pitohui) feathercollector/Shutterstock.com; p. 21 (stonefish) PRILL/Shutterstock.com; p. 21 (Iberian ribbed newt) Sergio Gutierrez Getino/Shutterstock.com; p. 21 (blue-ringed octopus) kaschibo/Shutterstock.com.

Printed in the United States of America

CPSIA compliance information: Batch #CS19GS: For further information contact Gareth Stevens, New York, New York at 1-800-542-2595.

CONTENTS

Words in the glossary appear in **bold** type the first time they are used in the text.

WARNING!

Stop! Are you sure you want to read this book? It's bursting with facts about beasts that can poison you. Just one sting, **stab**, or bite could be deadly.

If you're sure you want to continue, put on some thick gloves to guard your skin and turn the page. You're about to read about some awesome animals that have totally toxic **adaptations**. These adaptations help them stay alive but mean death for other creatures!

FACTS FOR THE FEARLESS

AFRICANIZED HONEYBEES ARE CALLED KILLER BEES. ONE STING ISN'T DEADLY, BUT HUNDREDS OF BEES MAY ATTACK AT THE SAME TIME. TOGETHER, THEY CAN KILL!

Many animals, including bees, have a bit of **venom**. That's why beestings hurt!

THE TERRIBLE TAIPAN

Wow, you turned the page! You really are brave. But your bravery has brought you face-to-face with one of the world's scariest snakes: the inland taipan. It has the deadliest venom of any snake on Earth!

When the inland taipan is ready to strike, its body forms an S-shape. As quick as lightning, it bites its prey. The venom travels from the snake's body through its sharp teeth, called fangs, and into the other animal. It can kill in less than an hour!

FACTS FOR THE FEARLESS

ONE BITE FROM AN INLAND TAIPAN CARRIES ENOUGH VENOM TO KILL 100 PEOPLE!

AUSTRALIA

INLAND TAIPAN RANGE

Luckily, the inland taipan lives in parts of Australia where few people live. Phew!

THE BEASTLY BOX JELLYFISH

You might think escaping into the ocean is a way to hide from dangerous animals. Wrong! Off the Australian coast and in the warm waters of the Indian and Pacific Oceans swims a strange-looking creature called the box jellyfish.

This jellyfish's body looks like a box with long, thin arms called tentacles streaming from it. Each tentacle can be 10 feet (3 m) long and has tiny body parts that work somewhat like poison darts. If stung by a box jellyfish, a person can die from its venom in just minutes!

FACTS FOR THE FEARLESS

UNLIKE OTHER KINDS OF JELLYFISH THAT TRAVEL WITH THE CURRENT, BOX JELLYFISH CAN SWIM.

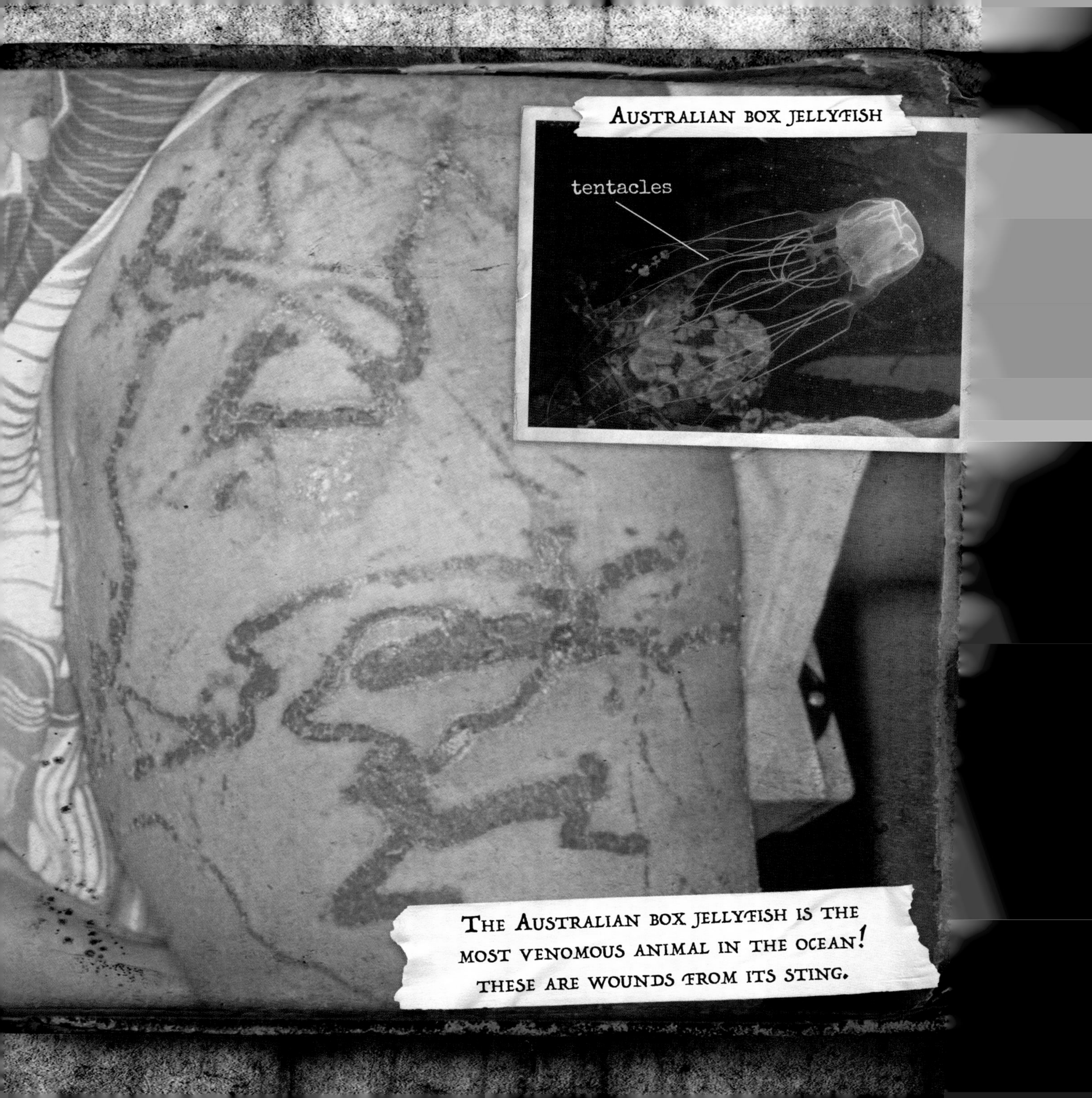
Australian box jellyfish
tentacles
The Australian box jellyfish is the most venomous animal in the ocean! These are wounds from its sting.

THE FRIGHTENING FROG

You escaped the jellyfish—good work! Now you're in a colorful rain forest in South America. Out of the corner of your eye, you spy the cutest little yellow frog. It looks like a toy. It's a golden poison dart frog!

Don't touch it! These frogs make a poison that oozes from their colorful skin. Just one golden poison dart frog makes enough poison to kill 10 people. Scientists think the frogs make the poison using **chemicals** in the bugs they eat.

FACTS FOR THE FEARLESS

WHAT'S THE DIFFERENCE BETWEEN VENOMOUS AND POISONOUS ANIMALS? VENOMOUS ANIMALS STING, BITE, OR STAB TO **INJECT** THEIR POISON, WHILE POISONOUS ANIMALS NEED TO BE EATEN OR TOUCHED FOR THEIR POISON'S EFFECTS TO BE FELT.

Poison dart frogs removed from the rain forest—and away from the bugs they eat—stop making poison!

THE WICKED WANDERING SPIDER

Eek! The rain forest floor is covered in spiders! They belong to a group of wandering spiders called *Phoneutria*. That means "murderess" in Greek! Perhaps the scariest spiders in this group are the Brazilian wandering spiders.

These **arachnids** don't build webs to catch prey. They hunt and use their venom to kill bugs, other spiders, and small animals. When people are bitten, they feel their skin burning. Their heart may beat faster or slower, and they might feel sick. The spiders rarely inject enough venom to kill people, though.

FACTS FOR THE FEARLESS

IN 2013, A FAMILY IN ENGLAND BROUGHT BANANAS—AND THE EGGS OF BRAZILIAN WANDERING SPIDERS—HOME FROM THE STORE! THE FAMILY HAD TO MOVE OUT OF THEIR HOUSE UNTIL THE DEADLY SPIDERS WERE REMOVED.

Brazilian wandering spiders are also called banana spiders because they're often found on banana leaves. This is one in a defense pose.

THE SCARY SCORPION

Let's run out of the rain forest and into the desert. What could poison you there? A scorpion could! The bark scorpion is an arachnid that lives in the southwestern United States. It's the only scorpion in North America that could kill you!

Like all scorpions, the bark scorpion has a tail that ends with a hollow stinger. When the scorpion stings, venom travels through the stinger into the other animal. Luckily, people who are stung by these arachnids can be saved by **antivenin**!

FACTS FOR THE FEARLESS

BARK SCORPIONS ARE THE ONLY SCORPIONS THAT CLIMB. THEY SOMETIMES CLIMB INTO PEOPLE'S HOMES!

Adult bark scorpions can be up to 3 inches (7.6 cm) long.

Bark scorpions are nocturnal, which means they only hunt at night. Their prey includes bugs, spiders, and other scorpions. Bark scorpions glow under certain kinds of light.

THE HIDEOUS GILA MONSTER

As you leave the desert, make sure you don't run into a Gila (HEE-lah) monster! Although this lizard is called a "monster," it's only about 20 inches (51 cm) long. Luckily, it's not very fast. Instead, it sneaks up on its prey and bites. It hangs on, letting its venom ooze into the animal's body.

The Gila monster's bite is very painful, but it doesn't usually kill humans. However, the Gila monster's venom easily kills young birds, frogs, mice, lizards, worms, bugs, and other small creatures.

FACTS FOR THE FEARLESS

THE GILA MONSTER AND THE MEXICAN BEADED LIZARD ARE THE ONLY VENOMOUS LIZARDS ON EARTH.

Gila monster with prey
Gila monsters can't see well.
They use their tongue to smell prey!

THE HORRIBLE HORNET

Maybe you think the skies are safe from toxic creatures. But that's not true! Hornets and bees are bugs to fear in some places. The Asian giant hornet, for example, can be as long as your thumb. Females of this species have stingers that are 1/4 inch (6 mm) long.

Asian giant hornets attack when people **disturb** their nests. Their venom can kill, especially if someone is stung many times. Running is a bad idea, though. These hornets will chase people!

FACTS FOR THE FEARLESS

ASIAN GIANT HORNETS ARE DRAWN TO HUMAN SWEAT AND SWEET SMELLS.

Asian giant hornets eat bees and wasps.

YOU SURVIVED!

You survived the beasts of this book! You avoided being stung, stabbed, bitten, or poisoned in any way. Now you know there are some really poisonous creatures out there. You don't have to be afraid, though. Scientists have discovered antivenins to save people from the deadly effects of many kinds of venom.

Besides, most animals aren't out to get you. They fear people and only use their deadly adaptations to protect themselves and their young. Still, it's best to give wild animals plenty of space!

FACTS FOR THE FEARLESS

THERE ARE SOME VENOMOUS **MAMMALS**, TOO. THE PLATYPUS IS ONE EXAMPLE!

More Totally Toxic Beasts

Platypus

Males have sharp venomous points on their back feet for protection.

Hawksbill Turtle

Toxic **algae** makes its meat poisonous to eat.

Hooded Pitohui

Poison in its skin and feathers protect it from predators.

Stonefish

Its **spines** can inject venom into predators.

Iberian Ribbed Newt

Its ribs can stick out to inject venom into predators.

Blue-Ringed Octopus

Its venomous bite can kill a human.

The best protection against these and other dangerous beasts is to arm yourself with knowledge! That way, you'll know where they live and how to avoid them.

GLOSSARY

adaptation: a change in a type of animal that makes it better able to live in its surroundings

algae: plantlike living things that are mostly found in water

antivenin: something that works against the effects of a venom

arachnid: one of a large class of small animals that includes spiders, scorpions, ticks, daddy longlegs, and mites

chemical: matter that can be mixed with other matter to cause changes

disturb: to bother or upset

inject: to use a sharp body part to force venom into an animal's body

mammal: a warm-blooded animal that has a backbone and hair, breathes air, and feeds milk to its young

spine: one of many stiff, pointed parts growing from an animal

stab: to quickly push a pointed object into or toward someone or something

venom: something an animal makes in its body that can harm other animals

For More Information

BOOKS

Cusick, Dawn. *Animals That Make Me Say Look Out!* Watertown, MA: Charlesbridge, 2016.

Meister, Cari. *Venomous Animals.* Minneapolis, MN: Pogo, 2016.

Sullivan, Laura L. *The Box Jellyfish.* New York, NY: Cavendish Square Publishing, 2017.

WEBSITES

Poison Frog
kids.sandiegozoo.org/animals/poison-frog
See more pictures of this colorful—and poisonous—frog!

Wacky Weekend: Deadly Creatures
kids.nationalgeographic.com/explore/wacky-weekend/deadly-creatures/
Check out some amazing facts and photos of creepy creatures.

Publisher's note to educators and parents: Our editors have carefully reviewed these websites to ensure that they are suitable for students. Many websites change frequently, however, and we cannot guarantee that a site's future contents will continue to meet our high standards of quality and educational value. Be advised that students should be closely supervised whenever they access the internet.

INDEX